Read for a Better World™

RAYS

A First Look

PERCY LEED

GRL Consultants,
Diane Craig and Monica Marx,
Certified Literacy Specialists

Lerner Publications ◆ Minneapolis

Educator Toolbox

Reading books is a great way for kids to express what they're interested in. Before reading this title, ask the reader these questions:

> What do you think this book is about? Look at the cover for clues.
>
> What do you already know about rays?
>
> What do you want to learn about rays?

Let's Read Together

Encourage the reader to use the pictures to understand the text.

Point out when the reader successfully sounds out a word.

Praise the reader for recognizing sight words such as *they* and *are*.

TABLE OF CONTENTS

Rays

Rays are fish. They live in warm ocean waters.

Rays have two large fins. Rays move their fins to swim.

Rays have thin tails
that can sting.
Rays sting sharks.

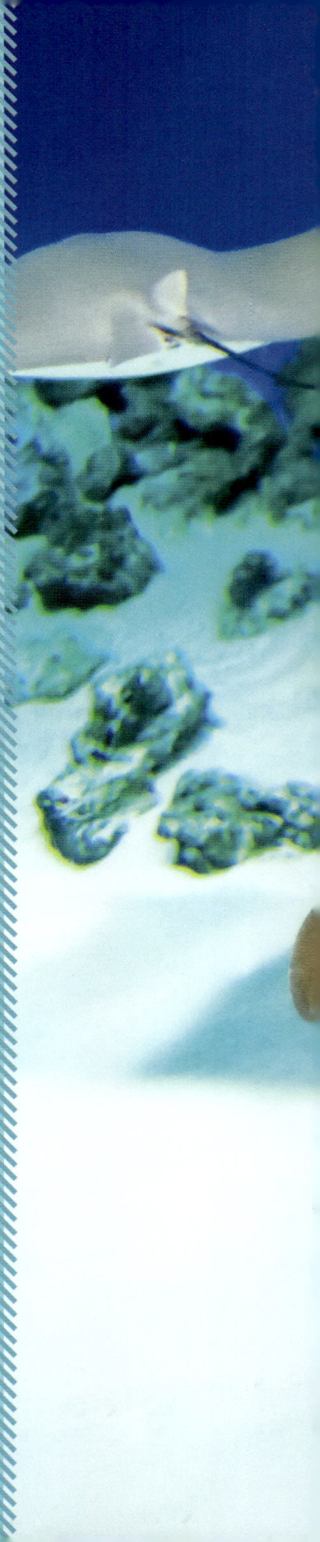

Rays hide in sand. Their colors match the sand.

How might a ray's color help it stay safe?

A ray has two eyes on the top of its body. Its mouth is on the bottom.

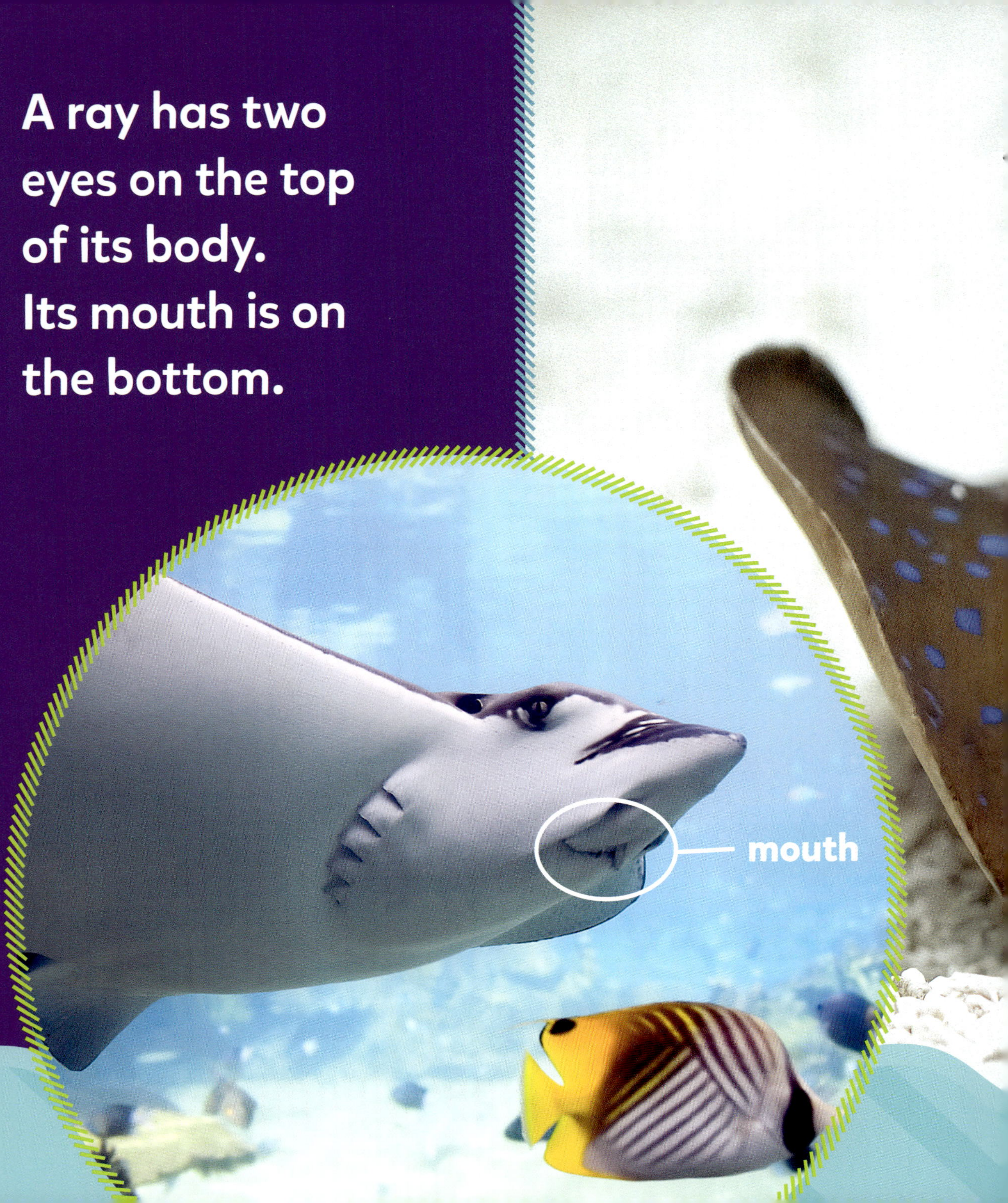

Why would it be helpful to have the eyes on top?

Rays use their strong teeth to eat.
They can break open clam shells.

teeth

Baby rays are called pups.

Mother rays have up to six pups at a time.

Some rays
live alone.
Others swim in
groups called
schools.

Why might some rays swim in groups?

Some rays jump high out of the water.

You Connect!

What is something you like about rays?

Have you ever seen a ray at an aquarium?

What other ocean animals do you know about?

STEM Snapshot

Encourage students to think and ask questions like scientists. Ask the reader:

What is something you learned about rays?

What is something you noticed in the photos of rays?

What is something you still want to learn about rays?

Photo Glossary

fin

pup

school

teeth

Learn More

Bodden, Valerie. *Rays*. Mankato, MN: Creative Education, 2022.

Leaf, Christina. *Stingrays*. Minneapolis: Bellwether Media, 2021.

Szymanski, Jennifer. *National Geographic Readers: In the Ocean*. Washington, DC: National Geographic Kids, 2018.

Index

Photo Acknowledgments

The images in this book are used with the permission of: © aapsky/iStockphoto, pp.8–9; © Andreas Polz/Shutterstock Images, p. 16; © Antagain/iStockphoto, pp. 12–13; © busypix/iStockphoto, pp. 18–19, 23 (school); © cdascher/iStockphoto, pp. 6–7, 23 (fin); © EXTREME-PHOTOGRAPHER/iStockphoto, pp. 17, 23 (pup); © Pascal Vosicki/Shutterstock Images, pp. 10–11; © qldian/iStockphoto, p. 12; © Reimar/Shutterstock Images, pp. 14–15, 23 (teeth); © renacal1/iStockphoto, p. 20; © User10095428_393/iStockphoto, pp. 4–5; © Wan Yong Chong/iStockphoto, p. 18.

Cover Photo: © cinoby/iStockphoto

Design Elements: © Mighty Media, Inc.

Lerner Publications Company
An imprint of Lerner Publishing Group, Inc.
241 First Avenue North
Minneapolis, MN 55401 USA

For reading levels and more information, look up this title at www.lernerbooks.com.

Main body text set in Mikado a Medium.
Typeface provided by Hannes von Doehren.

Library of Congress Cataloging-in-Publication Data

Names: Leed, Percy, 1968- author.
Title: Rays : a first look / Percy Leed ; GRL consultants, Diane Craig and Monica Marx, Certified Literacy Specialists.
Description: Minneapolis : Lerner Publications, [2023] | Series: Read about ocean animals (read for a better world) | Includes bibliographical references and index. | Audience: Ages 5-8 | Audience: Grades K-1 | Summary: "Rays use their two large fins to swim in warm ocean waters. Simple text and engaging photos introduce readers to these interesting ocean animals"— Provided by publisher.
Identifiers: LCCN 2021051290 (print) | LCCN 2021051291 (ebook) | ISBN 9781728459134 (library binding) | ISBN 9781728464176 (paperback) | ISBN 9781728461724 (ebook)
Subjects: LCSH: Rays (Fishes)—Juvenile literature.
Classification: LCC QL638.8 .L44 2023 (print) | LCC QL638.8 (ebook) | DDC 597.3/5—dc23/eng/20211021

LC record available at https://lccn.loc.gov/2021051290
LC ebook record available at https://lccn.loc.gov/2021051291

Manufactured in the United States of America
1 - CG - 7/15/22